BREAK-22

A H FITZSIMONS

Copyright © 2025 by A H FitzSimons

The right of A H FitzSimons to be identified as the author of this work has been asserted by him in accordance with the Copyright, Designs and Patents Act 1988

All rights reserved. No part of this publication may be reproduced, stored in a retrieval system, or transmitted in any other form or by any means, electronic, mechanical, photocopying, recording or otherwise without the prior written permission of the copyright holder

Published by Open Path Books 2025 ISBN 978-1-8383829-7-1
openpathbooks1@aol.com

Cover derived from the first edition of the novel
Catch-22 by Joseph Heller
Original cover design Paul Bacon

For Carole and Eva Thérèse

Acknowledgements

I thank the team: Ozzy Eyre, Major (Retd) Ian Johnson, Rubén Manso, Seán Pugh and Reno Sande. I would also like to thank Roni Ferguson, Dugald McCallum, John and Janice Miller and Bernward Wenzel for their ongoing support and encouragement.

Author's Note

Although I first read the book 'Banco' in the late 1970s, I only learnt of the controversy surrounding the author's alleged incarceration on Devil's Island when this book was being proofread. Consequently, the reference to Henri Charrière remains unaltered.

The pseudonym Reno Sande has been used for one of the team members owning to the nature of his violent past.

Contents

Introduction		13
I	Mission Impossible	17
II	The Plan	21
III	Special Protection Service: Night Moves	25
IV	Mireille: first extract from HK9	33
V	The Highway Nine	39
VI	First Video	41
VII	Mireille: second extract from HK9	47
VIII	T-shirts, Website and Hypocrisy	59
IX	Doubt	65
X	Second Video	71
XI	Breakthrough	77
Copyright Acknowledgements		85
About the Author		87

BREAK-22

Introduction

I first became concerned about global warming after watching *An Inconvenient Truth* in 2006. However, probably like thousands of others, I put it to the back of my mind and got on with life. But over time I became increasingly aware of the impending disaster facing mankind. More worryingly, I became aware of the lack of concern shown about this by the vast majority of the population. Eventually I felt I had a responsibility to do at least something to try to change public opinion. It was very clear that a level of denial existed. As time passed I realized a hard truth: we needed extreme measures to resolve the climate crisis, but world leaders and governments were not going to take these measures until the public were ready to accept them; and that wouldn't occur until the situation became much worse. The Catch being that by the time that happens, we will have passed the point of no return and it won't matter what we do.

I saw my task was then about finding a way, at least in theory, to break the Catch: to find a way that governments could introduce extreme measures before it was too late. It seemed organizations concerned about global warming were talking about extinction, but no one was talking about the horrors that would take place when the major crops failed and law-and-order ended. Hardly surprising, as no one wants to talk about that because no one wants to hear about it. But however ugly it is, and uncomfortable it makes us feel, we need to hear about it. We need to understand that global warming, the end of law-and-order and horrific violence are inexorably linked.

I have used a mixture of recounting of real events, as they happened to me, as well as quotes from a novel, in which the main character, Mireille, is attempting to prevent the impending real-life consequences of global warming we all face, but we'd prefer to deny. My aim in writing this book, like Mireille's, is to tear down that wall of denial - and offer a solution. Mireille became my voice, and with it being fiction there were no limits to what she could say. The second of the two excerpts from the novel HK9 used in this book looks at the future Mireille sees unfolding. The first extract describes a psychologist drawing a parallel between denial of global warming and denial of serious illness.

In respect to my team, far more is written about Reno than any other member. This was deliberate, and it is not just because he transpires to be the key member of the team. Reno is a symbol of man's primordial instinct to survive, more specifically of his resilience - his ability to shrug off hardship and whatever trauma life throws at him.

We may think we are capable of enduring difficult times. We only have to look back a handful of years to lockdown to confirm this. We did it then, when the danger in the form of a deadly virus was upon us, so at a deep level we may think that we can adopt that same survival mindset if/when the danger from global warming is upon us. When the storms and rainfall and forest fires become so bad that we finally call upon our instinct to survive and adopt similar measures to those we adopted in lockdown, and acknowledge that this time, instead of months, we are probably going to have to endure years of discipline, deprivation, hardship and sacrifice.

It worked with covid-19 and it worked in the Second World War ... when we finally recognize the danger is upon us we act decisively, and we do what we have to do ... and we prevail.

But that will not work with global warming. When we finally recognize the danger is upon us it will be too late. The momentum of rising temperatures, forest fires and melting icecaps will be unstoppable, irrespective of what we do. We are rapidly approaching the point of no return. The danger is upon us now, but the vast majority of the public can't see it. Nor do they want to. They don't want to fight, and they'll wait until the last moment before they do. Whereas in Reno we have people who are from a time when fighting for survival was part of their everyday life. In him, and others like him, there is hope. We all possess the primordial instinct to survive, but we need to shake off the feeling that we are entitled to life to uncover it.

Reno, and those like him, expect to have to fight for life, and it was fitting that he would come up with an answer as to how to break the Catch.

It was when he did, that I decided to write this short book. What follows is not an academic study, nor is it a polished story. It is a true account describing my numerous attempts to break the Catch, by breaking the denial. How each attempt failed, and the hypocrisy and greed I encountered along the way. Until finally, out of the blue, Reno came up with an answer that the rest of the team had completely overlooked.

I

I had a problem. The truth of it suddenly hit me at half-four on Friday morning. The time when the demons launch their main offensive. The incident took place the previous Saturday. Willie was having a leaving-do before heading back to Australia. After a sixteen month stay in hospital I lived in relative isolation in a flat in Edinburgh and rarely went to anything that involved people. But Reno Sande had insisted I attend, and he was going to pick me up en route. I'd met Reno whilst working nightclub doors in the mid-1990s. He was not a man you wanted to disappoint. Besides, despite my surface reluctance, this was one reunion I actually wanted to attend. Willie Kennaway, along with his brother Brian, were two of my oldest friends; we'd worked on nightclub doors together as far back as 1983. Brian having single-handedly saved me from getting a kicking from ten men in '84.

So, there I was in a bar, feeling out of place as usual, when I found myself talking to big Peter, former doorman, and currently a psychiatric nurse. Our conversation drifted on to the past, and how much we missed the '80s and '90s. I then started to describe a scene from the movie *Stand Up Guys* and how the characters played by two famous actors raided an old folks' home to take the third member of their gang, their former getaway driver, out for a series of adventures. I described all the sub-plots, the trip to the brothel, how they evaded two police cars in the high-performance car they'd stolen, then finding a hooker tied up in the boot ... who'd been given a hard time by the group of guys who'd kidnapped her. So, the three men decided to exact revenge on the group and, following directions, went to their hideout. When they arrived, the driver, who decided to remain in the car, commented to one of his friends that it was just like the old days. But his friend replied that it was better than the old days. The driver agreed that it was better, but couldn't understand why. His friend explained it was due to them being able to appreciate it this time, and the driver nodded in agreement as he suddenly grasped the truth of it.

However this time, as I recalled the conversation with Peter, I realized something was wrong. If it had been someone else I probably wouldn't have noticed. Although I had seen Peter on numerous occasions in the past, this was the first time I'd actually spoken to him. But instead of having a normal conversation I had spent twenty minutes describing the plot of a film to him, when there was no reason to do so.

When I asked the few friends I had about it, I discovered that what had taken place with Peter was normal for me - I invariably steered every conversation onto at least one film. I saw now how natural this was for me. I couldn't talk about anything else, because I didn't know anything else. I'd lived in relative isolation for the past 16 years - films had become my reality.

That wasn't the problem though, nor was it that I had a habit of pretending I was the main character in the films I'd seen, or that far too many of these films showed successful conclusions to impossible situations. The problem should have been that I had started to believe I could successfully complete impossible missions ... and that the mission I had chosen to complete was to stop global warming and resolve the climate crisis.

But that wasn't the problem either. What was, was that I had been reconditioned to the extent that I was now certain I could find a way to resolve the crisis, and having accepted the mission, I wasn't going to stop until I had.

Which was, of course, completely bonkers. But I didn't care if it was. I didn't care if somewhere along the line I had lost the plot completely and was now 100 per cent certifiable. When you consider that every time I looked in the mirror I saw Ethan Hunt, the main character of the *Mission Impossible* films, I almost certainly was.

II

One week later. Friday.

Sometimes when we dream, we know it's a dream. Often when this occurs we can exert a level of control over the dream, especially if we're near to waking; although the control is brief. This was not one of those dreams. I was looking at a mother and her young daughter. They were, in turn, looking out of a bedroom window at the rain. The child is asking a series of questions, wanting to know the reason it's raining and when it's going to stop. Her mother answers in a way the little girl will understand - that it's the man in the moon who is crying. The child asks if he stops crying, will it stop raining? The mother replies, giving reasons why he's crying, and why he can't stop ...

Her voice cuts off as she sees a small boat approaching. Men she doesn't recognize are on the deck. Pulling her child with her she cowers down underneath the window, and placing her index finger to her lips, signals for her child to be silent. The two of them go to the wardrobe at the side of the room. Sliding the door open the mother motions for her child to go to the far corner while she takes her place behind the door that she closes shut. At first, she hears nothing; then she hears the sound of men wading through the flooded rooms downstairs talking to each other. One asks if he should check upstairs. 'No,' replies the leader, 'I'll do it.' Then the heavy footsteps on the stairs being slowly ascended.

I had to do something. I had to help the mother and her child. I went to move towards the side of the bedroom door, but my body wouldn't respond. I was held in place by an invisible force. The footsteps were nearing the top of the stairs. This may have started as a dream, but a nightmare was unfolding. I gave it everything I had to move. My right leg jerked out in the bed as I woke from the dream. I didn't want it to unfold any further, with the sound of screaming as the mother and her child were discovered ...

There was no mercy in this nightmare. I looked to my right, the figure of 03:15 lit up the screen on the clock on the bedside table. Like so many long-forgotten dreams this one, on waking, was clear in my mind. My body wanted to go back to sleep. Sleep; it was so precious. I wrestled with the idea of getting up and putting pen to paper and recording the first part of the dream. But then I would be fully awake, and sleep would then be a day away. I had a busy day ahead of me – an Extinction Rebellion (XR) meeting; my first. I needed to be sharp for that.

You can rest at lunchtime.

I hauled myself out of bed. Two minutes later I was writing; the words flowed fast. I had to capture the dream before it slipped away; lost forever into the dead of night.

Sticking solely to the conversation between mother and daughter, what ended up on paper was a poem. I went back to bed, and for the next hour stared at the ceiling and the shadows in the room. Just as I thought, sleep was now lost for another day. At least I had the poem, and I knew exactly what I was going to do with it. Around four years ago I had started work on a novel. I could easily write it in to one of the chapters. The main character, Mireille, was an English Literature student, and, as of today, had a habit of writing poetry.

I planned to do a number of things in an attempt to break the wall of denial that surrounded global warming. I was determined, however, that the novel, via the voice of Mireille, would be the main battering ram that would bring down the wall. That was dependent, of course, on people reading it. The idea behind it being that, for all the warnings given out by XR, and others, they are going to have difficulty in describing the horrors that would take place when the major crops fail. But in a fictional book, Mireille would have no problem in letting everyone know exactly what was going to happen.

I was currently working on the final draft, rewriting most of her scenes. But there was still much to be done outside of the novel. It was going to be a big day. I had plans for today. I saw my first XR meeting as a recruiting opportunity. I had a crazy idea of forming a splinter group of XR called the Special Protection Service (SPS). This group had a total of five members: Ian, Ozzy, Rubén, Reno and myself, and our aim was to protect the planet using any means available.

Ian was a recently retired soldier, who joined up just before the school leaving age went up from 15 to 16. He was classed as the last of the boy soldiers. His Army career of 40 years began in 3 Para before transfer to the PT Corps, ending up as a Warrant Officer Class One. After a successful application for a commission, three further promotions followed, and he retired as a Major in what used to be the Education Corps. His final posting had him lecturing officers in Military History at Edinburgh Castle. In addition to his knowledge and experience of the military (which I was using to full effect in the novel) Ian also had a degree in English Lit, making him my first choice for proof reading.

Ozzy, although I had never met him, had worked with me on numerous projects relating to cancer. With what must be over a thousand emails taking place between us over the years, we had got to know each other quite well. Ozzy was well versed on computers, compiling videos and voice-overs. In things relating to the use of new technology, if he didn't know that much about it, he would research the subject until he did.

Rubén was the key member of the team. A senior scientist who specialised in forestry. We were all deeply concerned about climate change, but it was clear that Rubén worried about it the most. In addition to his vast knowledge, he was also a professional editor. As a result of his work on earlier drafts of the novel, I was now re-writing the second part of it, the part that focused on Mireille.

Reno was the final member of the group, and his particular set of skills would be crucial in the task I had planned to describe at the XR group meeting.

None of the team knew anything of my idea to form a splinter group called the Special Protection Service. I needed to recruit a lot more members before that happened.

III

I was running late, standing on Dundee Street looking for a taxi. Despite it being half-five on Friday afternoon the place was deserted. It felt odd, like I was on the set of an episode of the 1960s series *The Avengers,* where they employed two or three extras at most. Even in the distance I couldn't see any movement. Where was everyone?

Time passed. Another five minutes and I'd be late for the meeting. XR held this meeting every fortnight and I would be a new addition to the group so I should be on time. But if I was a few minutes late, would it matter? No. Not if I got there in the next thirty minutes. I really wanted to go to this meeting; my attendance was part of a plan. I needed another three people to pull it off. First stage of the plan was to recruit those three today.

I intended to hire a transit size van. Load the back with bags of sugar, bottles of water, funnels, pliers, screwdrivers, chisels. We would carry with us several hundred leaflets. The leaflets would read:

Your car has been targeted by the Special Protection Service – a splinter group of XR dedicated to, amongst other things, taking all petrol and diesel vehicles off the road. We have been driving around Edinburgh tonight in a van loaded with tools specifically designed to remove locked petrol caps, together with numerous bags of sugar, bottles of water and funnels.

We would then go on to say that the SPS has not touched their vehicle, then listing the environmental benefits if they traded their car in for an electric vehicle. Then going on to say that if we continue on our current course the major crops will fail. The worldwide food shortage that follows will be so extreme that law and order will end. The text will finish with the question: what will happen to the children then?

I didn't believe many car owners would read that paragraph though. As soon as they read about taking petrol and diesel cars off the road, sugar and funnels, they will assume that their car will now be off the road until their fuel tank is drained and cleaned. And if they'd turned the key in the ignition before reading the leaflet the carburettor or fuel injector would also need to be cleaned. Their first reaction would be anger and a phone call to the police.

We would set off around midnight, and finish by six. On each vehicle we 'hit' we would leave the leaflet sticking out of the thin gap in the panel that lies over the fuel tank.

To avoid any potential trouble with the Police, but mainly to get publicity, we would leave a letter at the Scotsman offices before setting out. The letter would have specific instructions that it was not to be opened before 6am, at which time it was to be handed to the first journalist on duty. We would also place all the sugar, bottles of water, tools and funnels in a large chest, secured by a padlock, the key to which would be left at my flat.

Reno would give the instructions to the security guard at the Scotsman offices. That would be enough to ensure the letter would not be read before we'd finished.

He would also be on hand when the leaflets were delivered, to ensure the safety of those placing them beside the fuel caps.

I had no idea at this time that when it came to the task of breaking the Catch, it would be Reno – who had little interest in global warming – who would prove to be the key member of the team. The irony was inescapable, especially considering he left school at 16 with no qualifications whatsoever, compared to the number of degrees held by the rest of the team.

Reno belonged to a particular group. We know them. There are not so many of them around these days, but occasionally we catch sight of one. We know them as throwbacks; men who should have been born in an earlier century. The mid-1800s was their time, when boys from working class families went through puberty deep underground, digging coal or enduring the white-hot heat from the steel furnaces. These boys were from strong stock; their ancestors having earned their living from hard labour, generation after generation to the point that it was in their DNA. They say there's no such thing as a pure mesomorph although these boys, and men, were not far off it – their shoulders broad, their hips narrow – but their bones were thick and solid, typical of an endomorph. These were bodies made to withstand back-breaking work in the worst conditions. Rising at four, after a hearty breakfast of porridge or bread and dripping they were off to work, walking miles to the yard or pit, their shift starting at six. By the age of twelve their days of childhood and puberty were long past, their skin as tough as leather, their hands calloused. By fifteen they were hard drinkers and had slept with half-a-dozen women. In today's world with the shipbuilding and mining industries in recession, and school compulsory until 16, these throwbacks, who had such physical superiority over others, would dominate school sports day. They would also typically be the school bullies.

Reno, was not just a throwback, but an extreme example. By the age of fifteen he had lost count of the women he'd slept with. Struggling to cope with the physical constraints placed on him by the modern world, he looked for ways to vent his frustration. As a rule, bullies would pick on boys who were incapable of fighting back. Reno, though, picked on the school bullies. He didn't waste time drawing them into a fight; he simply attacked them on sight. After beating so many of them senseless, word spread to the point where bullying became nonexistent at his school. Now he had to find another outlet for his fury. He managed to secure employment as cloakroom attendant at his local nightclub. At fifteen that was the only way he could officially work in licensed premises. Yet he was rarely seen in the cloakroom; instead he monitored the inside of the club for trouble in his unofficial capacity as nightclub doorman, aka bouncer. The day he turned sixteen he was moved from working the inside, to working the door. Often, a group of guys who had been refused entry would hang around outside directing their anger at the doormen, insulting them and attempting to goad them into a fight. When Reno was on the door all that ended. He would calmly step away from the main entrance and walk up to the biggest and loudest of the group – what exactly Reno said to him no one knew. But the group moved on. The Alpha had seen, and heard, something that frightened him. When you met Reno, the menace was there for all to see; both Ian and Rubén picked up on it within seconds of meeting him. Think Falconetti from *Rich Man, Poor Man*. Physically they were similar, but their voices were miles apart: Falconetti's voice was deep and threatening but Reno's voice was much deeper; it came from the depths of hell. Add to it the broad Scottish accent, and whoever he was talking to would realize this was someone, something, they had never encountered before ... and that scared them.

 Few men would be stupid enough to face him on their own, but sometimes a group of men would have a go. On one occasion he was attacked by such a group. One of them beat

Reno repeatedly over the head with a baseball bat. When they left, a battered and bloody Reno picked himself up and walked to his girlfriend's, collapsing on the sofa. First thing in the morning his girlfriend called in a doctor, fearful of the blood and multiple cuts and bruises on the man asleep on her sofa, whom she barely recognized. An hour later the doorbell rang waking Reno who stumbled to the front door. Standing on the doorstep, the doctor asked the man in front of him why he had a cushion stuck to the back of his head. When cleaning the wound, the doctor commented that the cushion had saved his life, if it hadn't sealed the wound he would have bled to death in his sleep.

More recently, in his fifties, he confronted a group of five young men at a caravan park. Reno was unaware there was a sixth man, who came up behind him and smashed a vodka bottle over his head. Reno went down, whereupon all six laid into him with savage kicks to his head and body. After a full minute, one of them, clearly the Alpha, shouted 'that's enough!'. The group dispersed, Reno staggered to his feet and walked home. On the rare occasions when he took a beating from a group of men, it was always the same: Reno would stand up, dust himself down, and carry on as if nothing had happened. All that was missing was the Peter Gunn theme from the film *The Blues Brothers*. This time, when he arrived home, the woman he was living with didn't recognize him; his face so swollen it resembled a watermelon. Yet despite the duration and savagery of the beating, none of his bones were broken. A normal man would have multiple fractures, and at least a major concussion, if not permanent brain damage; assuming, of course, that he survived.

Reno had been shot, stabbed and beaten over the head with almost every makeshift weapon known to man. His face and head carried the lumps and scars of those beatings, yet he remained indomitable, and as indestructible, as ever.

    His violent background made him the ideal person for this job. I would rent an electric van and while I drove, the three 'runners' would place the leaflets. Reno would follow up behind walking along the pavement ready to intercept anyone who posed a threat to the runners. He wouldn't have to get physical; all he needed to do was shout at any vehicle owner who was running towards the van – letting him know how stupid he was for not reading the rest of the leaflet and that his vehicle hadn't been touched. Hearing that voice and seeing Reno, and the fury and menace in his face, would stop them in their tracks and they would read the rest of the leaflet. Some might go inside and start to phone the police, then hang up before the call was answered, when they realized that no threat had been made. In fact, their first instincts were correct, the threat had been made, not in the words used, but in the tone and facial expression. And that, whichever way you looked at it, would not be viewed by a police officer, or a court, as a breach of the peace, let alone threatening behaviour.

    The way the leaflet was worded was another matter, but I felt all the important angles were covered. It was a good plan, though it would work better if we targeted every major city in the UK on the same night. But as it was now over half-an-hour after the XR meeting started, and still no sign of a taxi, it was looking like I wouldn't be recruiting anyone today. At least the meetings were held every fortnight. I consoled myself with that thought, as I walked back to the flat.

    Ten days later I called the venue where the meetings take place to confirm the start time on Friday. I was to learn that the online information was not up to date – XR meetings no longer took place at this venue.

As the weeks passed, I searched in vain for XR meetings in Edinburgh. After a few months my idea of flyers in fuel caps, and the creation of the SPS, faded away as I focused more and more on the novel.

IV

I had found the perfect spot in the novel for the poem. It was in the chapter where Helen Gleason, a senior manager at a secret government department, began to suspect that a student might be responsible for torturing and killing several industrialists. Helen knew the student, Mireille, had been exposed to a counter terrorist program. She wanted to know how the girl was still alive when every single test subject exposed to the program had either gone mad or committed suicide. The student wasn't just alive: the five-man team of former SAS operatives, who had been tasked with capturing her in Andorra, had failed to report in. If they were dead, as everyone seemed to believe they were, that would now be the second five-man Special Forces team the girl had wiped out. However, at the point I was rewriting in the novel, Helen was more concerned about the five men than anything else ...

The sound of Mozart's Piano Concerto No 21, from *Elvira Madigan,* softly embraced the living room of Gleason's flat. She stood at the bay window and looked out on to the night lights of Surrey. As tired as she was, she couldn't sleep. The thought of what had happened to the five men wouldn't leave her. After all, as she kept reminding herself, she was the one who had sent them on what was expected to be a capture mission.

    She went to her handbag, withdrew a piece of paper and read the words:

When will it stop raining?
When can we go downstairs?
When can I go outside?
When the man in the moon stops crying my child.

Why does he cry?
He cries because you can't play outside,
Because you can't run through his shadow,
Or see him dance with trees in the wind,
He cries for you,
Because you'll never see
The way he lit up the world at night.
But if he stops crying ... I would see him again?
He can't stop crying.
Like you?
Just like me, but his sorrow is greater,
He cries for you and every child,
Who will never run through his shadow,
Or see him dance with trees in the wind.

A simple poem, a child's questions about global warming. A distress call from the future, and reading it always raised the same question: was the group that was being hunted by Special Forces teams across the globe, not a group at all, but one person? One girl? Mireille Robertson?

It was too surreal, yet there were too many things that pointed to Mireille. The piece of paper Helen held in her hand that she'd found at the girl's home in Inverness, along with the books on global warming — far too many to be simple curiosity; this was more obsession. The timing: the killings of industrialists began two months after Mireille had been programmed with Hunter-Killer 9. The program itself — specifically designed to combat terrorism with greater terror. Helen had read the medical reports on the industrialists; they'd all suffered a violent and painful death. There was no one in the newly formed Extinction Rebellion, Greenpeace, or any of the organizations associated with saving the planet getting involved in violence, let alone doing something so ruthless. But Mireille, she'd been programmed to do things far worse.

All Helen's instincts screamed out to her that Mireille was top of the world's most wanted list. Dead or alive.

There was, however, one piece of the puzzle that didn't fit. What was she doing in Andorra? This region was mostly mountains and forests. There was no connection whatsoever to global warming. It made no sense.

Her being alive, didn't make sense either. Two years earlier, not long after the murders in Edinburgh, Helen had sought the opinion of Arthur, the dear old psychologist who came up with answers to seemingly unanswerable questions. Arthur would sit in his room, surrounded by ancient books, and ponder until he found a solution. He would have been replaced years ago, but he was much loved by the younger staff, so Human Resources had decided to keep him on for the morale of all, it seemed.

'Hello Helen,' Arthur stood up as she walked in to his room. He was probably the only man in the entire building who held on to the rapidly dying tradition of standing when a woman entered the room. But Helen also knew it gave him the excuse to walk to the large photograph on the wall, where he would look at it and say with pride, 'Have you seen my new car? Isn't she a beauty? All electric, charges up in two shakes of a lamb's tail.'

'I've heard that. It looks good,' Helen played along. 'I love the colour.'

'Are you still driving that diesel?'

'Yes, still have it.'

'You do know Helen, that you're in denial of global warming.'

She nodded in resignation to what was coming — Arthur's denial speech.

The psychologist began, 'Denial is a coping mechanism; it allows us to shut out things that we ordinarily would struggle with. Its existence allows us to function on a daily basis.'

Helen had heard these words a few months ago, but this speech was something that had to be suffered. An audience with Arthur came at a price ...

'When something happens, which is so traumatic that it might drive someone to suicide or insanity, we go into shock and the shock is frequently followed by denial, the mind shutting out all memory of it until such time as it is ready to accept parts of it. In this type of situation denial can be a very powerful survival mechanism. As it can be when we are unwell in hospital. When it comes to fighting illness belief is so important, and those who understand this will often deny an unfavourable prognosis and hold on to the belief that they will make a full recovery. Denial is allowing them to live each day free of fear and depression, they sleep better and the body's own resources work more efficiently. It doesn't mean they will survive, but if they can persevere with this approach it should increase their chances of survival.

'But when denial is employed at the wrong time it can be terribly dangerous. Often people go into denial of the symptoms before they see a doctor, convincing themselves that the difficulties they're experiencing will go away in time. In not starting treatment they allow disease to spread unchecked through their body.

'With global warming, denial allows us to sleep well and carry on our normal life, but in doing so, we continue doing things that will accelerate the warming as opposed to stopping it. Just like the people who are in denial of serious illness, we need to start on the right course of treatment. And the right course is to get rid of your diesel or petrol-fuelled car, stop travelling abroad and, well, I imagine you know the rest ... it's a long list.'

'As it happens, I am changing my car at the end of this year. I will be giving serious consideration to trading-in against a new electric car.'

'Oh well, at least you're thinking about it. But, while you're doing that also think on this: the main indicator of the level of denial of global warming is not the continued use of petrol and diesel-fuelled vehicles, or flights to holiday destinations, but the lack of concern by millions of parents about the future their children face.'

The psychologist allowed that to sink in for a few seconds before nodding, the signal that his lecture, and the game, had concluded. Arthur returned to sit behind his desk, Helen sat opposite him.

'Now, how can I help?' Arthur said.

'Mireille Robertson.'

'Well, I should have known ...

V

In November 2021, nine activists from Insulate Britain, an offshoot of Extinction Rebellion, were jailed for breaching an injunction banning protest activity on the M25. This was at a time when road travel was frequently disrupted by activists blocking major routes.

As the Highway Nine would be locked up over Christmas, I felt that they might want to read my novel, especially as it described the thought process and actions of a student who viewed anyone who abused the future of the planet as a terrorist. I expected, based on the many films I'd seen, that it would be easy enough to visit them, and even easier to contact them.

However, unlike the movies, contacting members of the Highway Nine directly proved to be near impossible. Each of them had a protector (a relative or close friend) and contact had to be with them. If they saw fit they would inform the member of your good wishes, but passing any literature on to them was forbidden by the prison service.

I didn't ask why. I assumed that those in power had had enough of the road blocks, and the ensuing gridlock, and were creating a deterrent by setting an example using the Highway Nine. By imprisoning them and restricting their contact with the outside world, the legal system was flexing its muscle, letting the public (and the Nine) know that the power lies with the authorities, and the next batch of protestors to appear in court for blocking a motorway would be treated just as harshly.

I spent a month communicating with the members' protectors. It was clear I wasn't going to get the book to any of the Nine, so it was pointless spending any more time on it. I dropped the whole thing. Besides, I realized that the novel I thought was done and dusted needed a couple of additional chapters. I also knew I needed to spend more time at the gym.

VI

I was leaving the gym, psyching myself up for the walk back to the flat. As I stepped outside into the sunshine and a wall of heat, I knew I wouldn't make it. I needed to sit down and rest for a while. I looked around for a suitable place and noticed Seán sitting on the low wall that ran to the left side of the entrance. On seeing me he asked how my session had gone; I answered that I had overdone it a little and sat down alongside him. We'd spoken briefly in the gym over the last few months, but this was the first time we engaged in an actual conversation. I'd seen Seán doing different things around the gym and surmised that he was there purely to get out of the house. He wouldn't be a pensioner, but he wasn't far off it. Old, but with life left in him. I guess the local gyms are where you would find men and women like him. It would be the place they would congregate and pretend they're still young – until the moment when they caught their reflection in the mirror. When I first started going to gyms you couldn't find a mirror; now they were everywhere.

Yet, I somehow knew that you would never have found young Seán in a gym. He had been neither city nor country boy; he would have been out at sea as soon as he left school, learning his trade. His voice and weathered features left me with the feeling that his youth had been spent sailing around the world. Later he would have spent his forties and fifties surviving in the harsh conditions of the North Sea. But he was no deckhand; I would put money on him being the captain of a deep-sea trawler. I imagined he was simply following in the footsteps of his ancestors. His grandfather would have captained one of the huge tall ships of old, and kept in his quarters a cat o' nine tails which Seán's great-grandfather had probably used ... and his grandfather had brandished a few times, to scare any of the crew who grumbled about work conditions. Not that many did in those days; they were hard men who climbed the masts in the midst of a hurricane, climbed carrying extra weight, their clothes soaked through. There were no safety lines, or safety measures. Many men would be lost on these journeys. Those who survived would go below at the end of their watch and sleep in their wet clothes – that being the only way to dry them.

Those chosen to lead these hard men almost always spoke with a voice that by itself commanded authority. And Seán was no exception. When he spoke in a level tone, you automatically paid attention as inherent in the gravel-like timbre and steel was an underlying threat. It was so subtle that only the unconscious would pick up on the warning of imminent danger that we all felt though didn't quite know why. In this respect he was the opposite of Reno. There was no mistaking the menace in Reno's voice: you knew the moment he opened his mouth you were in grave danger. With Seán the danger was no less, the difference being with Seán it was camouflaged – he was steel fist hidden in a velvet glove.

With that threat in mind, as opposed to asking him directly about his seafaring days, I steered the conversation around to global warming, telling him I was so concerned about the level of denial by the general public that I was going to make a film about it. It was obvious Seán was as concerned as I was about the denial. We both agreed that it wouldn't be that long before the major crops failed. I then shifted the conversation, bringing up the issue of overfishing due to fish becoming one of the last sources of protein. There was a pause in the conversation as I waited for him to reveal the true depth of his knowledge about the subject. Inexplicably, Seán moved on to another totally unrelated issue, as if he had no knowledge or interest in fishing. I wasn't buying that though, not for a second. I surmised that his days as a trawler captain had ended so badly that he would have good reason not to want to talk about it. Something terrible had occurred on one of his trips, something devastating. I had visions of a giant wave tossing a ferocious shark on to the deck of the ship, whereupon it had started to devour a crew member ... until Seán harpooned it from the bridge. Visions of Seán, as opposed to Sterling Hayden, walking to a gunfight at the end of *Terror in a Texas Town,* armed only with a long harpoon, flashed through my mind.

As much as I wanted to hear the details of Seán's adventures, it was clear they would never see the light of day; buried as they were in an ice-cold grave countless fathoms below the waves, lost in a storm ... lost at sea.

Saddened by this knowledge, I stood, explained that I needed to get back to work, and began the walk back to the flat. I had only taken one step when I heard Seán say 'When you finish the video, I'll do the music for it.'

I raced back to the flat in record time. I went online immediately, hurrying to find the information before the aftereffects of the workout kicked in and I began to feel unwell and had to lie flat.

By early evening I was able to think through what I'd learnt from the net. Seán was a composer of some note. I'd listened to enough to pick up on that. Aware that I had been so completely wrong about his past would have troubled me, but it was offset by the knowledge that I now had a soundtrack for the video ... that I hadn't even written yet. I did, however, have an idea of the footage to use on it, and as a bonus, the owner of the footage had the perfect voice to go with it.

I'd worked with Ozzy to help create a video for cancer patients years earlier. Ozzy had since developed his own website and had posted on it a video he had made of

Rottingdean beach. The footage was shot in the heart of winter, the beach mostly deserted. With dark jetties reaching out into a grey green sea and the cold white of the rock face as a backdrop, all topped off with a light grey blue sky, the footage possessed a dream-like quality. Ozzy had an eye for picking the right shots, and in the harshness of bleak winter, and the desolation, there was something quite brutal about his footage. Yet, the contrast between the greys, blues and greens gave it a beauty. There was nothing commercial about this beach in winter, the polar opposite of the typical seaside town beach. And that made it perfect for the video. I asked Ozzy about creating a video using his footage. He was delighted to use it, and to do the voiceover. With all the elements now in place, it allowed me to focus on writing the text.

It all moved pretty quickly: as soon as I finished the draft, Rubén edited it and Ian proofed it. Ozzy then did the voiceover. That's when things slowed down. I had to listen to the soundtrack a few times before I realized what was wrong. As great as Ozzy's voice was, it had been used to promote companies and their products for decades. His voice was perfect for advertising banks and building societies, and I was surprised that I hadn't ever heard it on TV. His voice did one thing above all – it evoked trust, and that was why I thought it would be perfect for this video. However, it had an upbeat, and optimistic edge, and as hard as Ozzy tried, those positive elements were near impossible to remove entirely. They were subtle, barely reaching the conscious; but they were there. Reluctantly, I asked Ozzy if he would mind if I tried recording the voiceover. I knew what his answer would be before I asked. Ozzy was a team player; he wanted the best video we could produce, be that using my voice, Ian's, Seán's or Rubén's (but definitely not Reno's). I spoke to Seán about it.

As luck would have it he had a film quality portable recorder, so we planned a day for the following week. He recorded my voiceover at the gym first thing in the morning when it was quiet. Seán now knew the final length so immediately began working on the soundtrack. Within a few days he had finished. He then laid the voiceover and soundtrack together, put it all on a flash drive and I posted it to Ozzy.

Three weeks later the film was on YouTube.

Two weeks after that all the contacts within the team had seen it and the number of views spluttered to a total stop at the figure of 33. A lot of effort had gone into this; and so far, that effort was wasted.

As neither I, nor other members of the team, knew of anyone involved in social media, it seemed clear that I needed to employ professional help. I went online to a company that promoted freelancers. You picked the freelancer from their website; deposited the funds for the job, the company held these funds in the escrow until the job was done; then you gave permission to release the funds and wrote a review.

In theory it should run smoothly.

I selected a young dynamic company that was listed in the top freelance companies available online. A company that had a multitude of glowing reviews. My instructions: to promote the video, increase views and comments on it, but under no circumstances to post false views.

They went to work and a bunch of views, about 300, appeared. So, I released the funds and wrote a positive review. I'd achieved what I wanted: to bring attention to the video. Some of the 300 would recommend the video to friends and, given enough time, the number of views would hopefully snowball. But that wasn't to be. The company employed to provide genuine views had done little to promote the video.

What they had done was exactly what I'd specifically asked them not to do. Two weeks later all 300 views disappeared. YouTube had deleted them due to them being false views.

At first I was raging, that this company had completely ignored my instructions and done their thing to ensure my funds would be released from the escrow and I'd post a favourable review. I made a formal complaint to the company who owned the website that promoted the freelancers. Although I knew the reviews were false, and who had posted them, I had no way of proving it. As time passed, I realized the company had, in fact, done me a favour. I had rushed every aspect of the video that I was involved in. It wasn't good. The text had been heavily influenced by the UK government breaking promises made at the last COP meeting, when they announced that new licences were available to drill for oil in the North Sea. That was relatively minor when compared to the breaking of other promises made by much larger countries. When you combine my poor text with my failure to adequately rehearse the voiceover, you had a video that frankly wasn't worth promoting.

I retreated into my isolated world, and resumed rewriting the novel.

VII

I was flying. The wind was at my back and I was unstoppable. I was running so fast now I couldn't hear the sound of my bare feet touching the wooden floor of the bridge.

At least, I thought it was a bridge. It was odd that I couldn't hear any sound coming from my feet hitting it. *I should hear sound. Am I running on a bridge? It had looked like a bridge when I approached it. I should check it's there. I need to know what I'm running on, don't I?*

No, just keep running.

But don't I have to know?

No, you have to keep running.

So I ran. Then, after a while I looked into the distance, but what I had seen earlier was no longer there. There was nothing there. I tried to feel what was beneath the soles of my feet, but I felt nothing.

Keep running.

Where am I? What am I running on?

It doesn't matter. Nothing matters except the run.

I have to check, to make sure ...

You don't have to do anything. Nothing matters here, just the run.

But I need to know. I need to make sense of this.

No, you only think you do. But it's not true, nothing is true: there is only the run, nothing else.

But what is the run if I can't see where I'm going ... if I can't see what I'm running on? I need to see what I'm running on.

Why do you need to know anything?

I don't know why, I just have to. I have to look now.

Another dream. I seemed to be having so many of them now.

At least this one wasn't a nightmare, but it was easy to recognize that the message contained within it wasn't good. I was losing – not just direction – but confidence in what I was trying to do. What was the point to this mission, if all I could come up with was hypothetical, and had no place in reality?

It was different for Mireille, she didn't have to find an answer, she could work away at eroding the wall of denial, and continue killing. But whilst doing that it was feasible for her to question the path she was taking. Then the dream would work in the novel. It felt right for her to doubt herself. At least it was a break from violent situations that she kept finding herself in.

Currently she was in Richmond, having gone to Helen's flat with the purpose of interrogating, then killing her. However, in questioning her, Mireille realized she had an ally. At the point I was now writing, the two women were driving towards the department ...

The city of London was a hive of activity around midnight. The Volvo sped past numerous black cabs as it made its way towards the department building.

The early part of the journey passed with Helen describing the protocol at the security desks and the layout of the fifth floor.

When the driver felt she had adequately prepared the girl she said, 'I was in charge of the team that searched your home.'

'And?'

' ... *He cries for you and every child, who will never run through his shadow, or see him dance with trees in the wind.*'

Mireille looked at Helen, 'You made your decision based on a doodle?'

'A poem.'

'That was hardly a poem.'

'I also found the books.'

'I read. I was studying literature.'

'There were too many books about climate change, and all together far too many coincidences. I understand you're working to create a deterrent, but you must know that everything you've done is being kept quiet, the media know nothing about any of the fifteen murders.'

A long silence followed, with Helen negotiating several roundabouts before the girl in the passenger seat said, 'Twenty-eight. It appears some of your security services and government agencies do not want to share information after all.'

'Where ... what countries?'

'You should find out on Tuesday. I had always planned to notify the press from London. I mailed the letters as soon as I arrived here.'

'Do you really think twenty-eight will be enough?'

'The research on deterrent shows that increasing the likelihood of being caught is far more effective than increasing the punishment. But the research is flawed: it looks at longer prison sentences being given out, not terrible, premature deaths. Fear is the real deterrent. It's not as if I am asking them to close their businesses. I am asking that they take measures to protect the environment.'

'Asking? You're hardly asking. You were programmed to do this, but you can't keep travelling the world, hunting down men and women who care nothing about the planet and killing them. You will be caught and killed. Any deterrent you've created will then cease to exist.'

Mireille said quietly, 'They care nothing about the children.'

'The children?'

A police car pulled out on to the road in front of them.

They drove behind it for two minutes before it indicated, then turned right.

Mireille broke the silence, 'You're like the rest of them, aren't you?'

'The rest of who?'

'In a far-off galaxy a teacher is asking questions of her class. "Planet 745 in sector nine of the fourth quadrant, what was the reason for the end to life?"

'A student at the back calls out, "Denial."

'"Denial of what?" the teacher asks.

'"Climate change and its implications."

'"Good. And do you know the root cause of the denial? Anyone? Okay, the human race had, over time, forgotten one of the fundamental rules to life. And it's the same rule for every life form on every planet in the universe. Survival requires a fight. The human race thought there was no need to fight climate change, fight those making money out of it, fight their own selfish habits that ultimately boosted the whole thing. They started taking life for granted. In their own words — they sowed the seed and left their children to reap the whirlwind."'

They waited at traffic lights. Helen wanted to say something, but again it was the girl who spoke.

'The priority isn't enforcing measures to reduce carbon emissions: it is bringing down the wall of denial. Once that wall is down, there will be no need to enforce anything, everyone will be onboard and doing everything they can to help. People won't drive cars that run off petrol or diesel. Houses won't be built without proper insulation. Air travel will effectively end except for essential and emergency services. Instead of going abroad on holiday, people will volunteer to plant trees, or build irrigation and drainage systems to prevent flooding. Everyone will work together to ensure survival, but the wall of denial has to come down first.'

'Okay, tell me, why do you think the world is in denial?'

'We've been told by the experts, we know the problems. Overpopulation being one of the biggest, birth rates may be falling — but not quickly enough — they need to literally plummet. Car manufacturers are still designing and building petrol and diesel driven cars, and the public are still buying them. People know the damage air travel makes to the environment, yet they still flock to the airports during the holidays. We talk about planting millions of trees, cutting carbon emissions to zero by 2050. There's far too much talk and minimal action. It seems the world has suddenly and conveniently forgotten the meaning of the verb 'accelerate', and how difficult it is to stop something when it gathers momentum. You can stop this car in a second, but an oil tanker takes twenty minutes to stop. Think size, and how infinitesimally tiny an oil tanker is, compared to the planet. The ice caps have started melting, how long do you think it's going to take to stop that process? I don't know of any trees that grow when they are on fire, or have been swept into a river, and 2050 will be far too late, as will 2040, and 2030. By 2030 we will have passed the point of no return.'

'Alright. So there's a wall of denial, how do you breach it?'

'Jeanne Achterberg.'

'What ... who?'

'Never mind. One route is through the visual cortex, the area of the brain responsible for processing images. The visual cortex can't tell the difference between what is vividly imagined and reality. Clarity of image and repetition being the key factors.'

'You've lost me.'

'It's the repetition that's important. When I watched the

film *An Inconvenient Truth* the message went right in, but then, after a few days, it started to fade. Yet, every time I saw an image that related to the original message in the film, I felt the truth of it again. I didn't need to see the film again. I didn't need the context. The image on its own was enough.'

'You do know that the vast majority of the public won't watch a film on global warming.'

'Yes, but the military and the police will, they don't have a choice. The government made sure they had the police and the military behind them before taking on the Miners' Union in 1983. Thatcher knew it was going to be a long fight, she had everything set up in preparation for it.'

'Okay, let's assume you have every member of parliament, every police officer and member of the military watch at least one film. I admit, the imagery in these films is powerful. So, those in power, and those who enforce power are on board, but for how long? A few days, weeks, and to what end? What's the point if you can't make the public watch films?'

'Not films, no. But you can get the messages through by using powerful images and repetition. Achterberg researched this. One of the most powerful images is that of an emaciated polar bear. Others of beached whales that had become confused by changes to the ocean, terrified animals on fire running out of burning forests, drowned cattle, and sheep floating in lakes where fields used to be. There are thousands of genuine images that scream out the warnings of what's happening to the planet ... vast chunks of icebergs sliding in to the ocean, dry ground cracked with the heat of the sun where lakes used to be. Let people see the then and now photos, the deterioration, the trend.'

'But where are the public going to see these images?'

'In newspapers, on billboards, on television. In every

magazine, every TV channel, every cinema, every single film clip on the internet. Go in gently at first, so the public barely notices. The films need to be short anyway, just two or three seconds, short enough that there isn't time to switch off or change channel. Target those in charge of the Media. They will all have their legal responsibility spelt out to them: that they will have to incorporate a certain number of images or short films on a daily basis, spread out evenly. And, if they don't comply, they face huge fines, and if they don't pay the fines, they go to jail. No trial dates, no appeals, no delays.'

'Why don't you just lock them up in the Tower and starve them to death? What you're saying is far too extreme.'

'The government — all governments — need to be decisive. All the opposition parties need to stop their bickering, drop their agendas and unify on this, just like they do in war time.'

'But what laws are going to ... the Government will be thrown out. You're talking about Big Brother gone mad.'

'Yes, you're right, it is Big Brother, but protecting everyone in the country who is, in ignoring global warming, effectively committing suicide ... and dragging the rest of us down with them.'

The car swerved.

'You'd best watch the road,' Mireille said softly.

Helen shook her head, 'There is a twisted, almost perverse logic to what you've just said. If you're going to go that route, then you missed out prison wardens from the list of those who will have to watch the films. And, the Government preparation should probably include building a lot of new prisons, and initiating a major recruiting drive for prison wardens, police officers and the military.'

A few seconds passed in silence, before Helen added, 'And, you're forgetting something else — the billions around the world who don't have access to newspapers, billboards, television and the internet.'

'If you'd ever lived in the highlands of Scotland, you would know that there is a specially adapted lorry that drives from village to village. The lorry is a mobile cinema. In warmer climates, huge screens could be set up outside, and films shown to thousands at a time.'

'But we're back to the original stumbling block — you simply can't make people watch films on global warming.'

'Have the main film about something else entirely. Have a break in the middle where you show the country's heroes, actors, sportsmen and women, or the senior and well-respected figures from that area, talking about global warming and what the general public can do to help.'

'Fine, let's assume you are right. Why are *you* not trying to breach the wall of denial, promote your ideas instead of killing a few industrialists?'

'From everything I've seen and read, it seems that the risk of extinction is too big for the mind to handle. Yet people should be able to grasp the level of violence that will take place if the major crops fail. At the moment people don't associate global warming with extreme violence. Yet they should as the two are inexorably linked.'

'But that link will only go to you—'

'You're missing the point — what do you think I put in the letters to all those journalists?'

'So, you told them what you've just told me, okay, that makes sense.'

The girl nodded, 'Journalists look for headlines before they'll take the time to read through what you've written. The level of violence grabs their attention. They know there is a story there to start with. And in a lot of cases the journalists concerned will have children.'

Helen, remembering Arthur mentioning the children, said, 'Everyone keeps talking about the children? We all know they have the most to lose.'

'It will hit them hardest. They're not responsible for destroying the planet. They are the ones who are doing what they can to protect it. And they've finally found their voice through a girl from Sweden. And if people don't listen to her, if we, as a planet stay in denial, all the children will suffer terribly when we realize we've passed the point of no return and hell—.'

'So we're back to the breakdown of law and order?'

'Breakdown? No. Breakdown implies recovery. There will be no recovery. Law and order will end — period — and that's entirely different. When I talk about extreme violence, you're thinking of something else. You are like most of the population, naive when it comes to understanding violence. When law and order ends the things that will take place will make what I did to the industrialists, the three men in your flat, and what those men intended doing to you and your daughter seem tame in comparison. There's a film, *Irreversible*, it has a scene in a deserted subway that a lot of people were unable to watch, so they walked out of the cinema. Their minds just wouldn't go there; it was too disturbing, too shocking. The terrible truth is, that after we pass the point of no return, the acts in this film will become commonplace. It will be bedlam; and it will be irreversible.'

'It's just a film, Mireille, the people who walked out probably knew it was deliberate shock tactics by the director, and that it would never take place in real life.'

The girl looked out of her side window. Seconds passed in silence until she said, 'When I turned fifteen, I sat down with my grandfather and told him that I was becoming jaded with the daily unarmed combat training. The skill level that he had taken me to, after ten years, would be more than enough to handle any situation that came my way. He sat across from me and nodded, and said I was old enough to make my own decision, and that he was fine with it. That took me by surprise. Anyway, the following week we're watching a French film a friend of his had recommended. Neither of us had heard of it before. Five minutes in, Grandpapa takes a phone call and explains he has to go out for a while. I'm left watching the film on my own. I had no idea at that point that the whole thing was a set-up. An hour later, I knew. Not long after the film ended, Grandpapa returns and asks me if I watched the rest of the film. I say I have, and I want to restart combat training tomorrow. The way I say it, he recognizes I know he set the whole thing up to achieve just that. I remember clearly what he said then, because he was angry, and I'd never seen him angry before. "I am sorry, sweetheart, I am sorry you had to see that. But you need to know that what happened in that film happens in the real world. Similar incidents are taking place right now, all over the world. Right now! Never forget how cruel people can be."'

'And the film ... it was *Irreversible*?' Helen asked.

The girl nodded.

A long silence followed.

Helen said, 'Hunter-Killer 9—.'

'Stop! Haven't you heard a word I've said?' Mireille looked down and shook her head. 'The point of no return is looming, and you need to acknowledge that. And when you do, when the wall of denial comes down, you'll know because you will be indifferent to anything associated with the Hunter-Killer project. Your focus will be on what you can do to save the planet and your daughter.

'We are so predictable; we continue to live in Never-Never Land where it's okay to think we can wait until the last minute before we need to take action. We have so much experience of resolving issues at the eleventh hour that, with a bit more effort, there's nothing we can't fix. But there's a thing about momentum: you don't see it until it's too late.

'Right now, we have meetings where top professors and scientists with masses of qualifications and experience, stand up and give a detailed lecture on the damage we're doing to the planet. We don't need any more detail — it's simple: we have never experienced anything on this scale before, and with such a force of momentum. We don't know how fast things are going to deteriorate. We don't know how much momentum global warming has already built. We don't know where the point of no return is. So how can we hold back from taking decisive action? If there's a fire in a block of flats you don't talk about the details of how fast the fire is burning, and measure temperatures, or discuss how much of the building can be saved — you get everyone the hell out of there as fast as you can — and put the fire out. You only begin to address the rest after you know everyone is safe. We can't evacuate the planet, and the people who are spending billions on exploring space under the premise that we can, are delusional — there isn't time! If we keep talking about this problem it will continue to gather pace, and we won't be able to stop it. What will the governments do then? Say we're sorry, but we'd never experienced anything like this before, we underestimated so many things ... if only we'd started earlier. We are so sorry, but we don't want anyone to have to face the large gangs that are now running amok across the globe, so just go to your local pharmacy where suicide tablets are being distributed.

'It won't be survival of the fittest, Helen, it will be survival of the cruellest! And even they won't survive long. When things get so bad that we realize we have to act, it will be far too late. The fight for survival isn't tomorrow, it is today, it is now. And right now we need men and women to stand up and be counted; it won't take that many — the rest will follow. There are so many opportunities that are available, yet they're not being taken. So far, I only know of one example: a leading rock band that has stopped touring. But it is a start. The wall of denial comes down soon, and I mean soon, we unify and start acting decisively. The wall stays up too long, major crops fail, law and order ends, bedlam ensues and the spark that was the human race is extinguished forever. There's no grey area here, no middle ground. It's one or the other ... and we're running out of time.'

VIII

I had republished the novel – the third, and now hopefully last, version. As always when self-publishing, whether in paperback or eBook, you need to promote the book. T-shirts seemed to be a way to go when combined with a website and a targeted campaign on social media. I thought if I created the right image on the T-shirt with the right text, then the T-shirt could become a sought-after item.

I was being terribly naive. If I'd taken the time to think about it, I would have realized that the only times I had ever seen more than one of a specific image or writing on a Tee is when staff of a company or group/team were wearing it.

But I didn't think about it. Instead, I set out with a positive mind-set that I could come up with something so original that it would become, in movie terms, 'a cult classic'.

The first Tee detailed Mireille's fight and was more about the book than about global warming.

A nineteen-year-old student from the Scottish Highlands has discovered how to bring down the wall of denial and resolve the climate crisis. As she puts her plan into operation, Special Forces units from around the world are deployed to kill her.
She is fighting for survival.
Are you?

The second focused predominantly on the tag-line of the book: **If you're not afraid, you're in denial.**

By the time I came up with the third Tee I had had more time to think about it. It transpired to be the first of two T-shirt's where the message was clear:

Denial *n.* A refusal to accept the truth or reality of something. The main indicator of the level of denial of global warming is not the continued use of petrol and diesel-fuelled vehicles or flights to holiday destinations but the lack of concern by millions of parents about the future their children face.

The fourth and final T shirt felt somehow less powerful:

> Governments are stalling
> Ice caps are melting
> Water levels are rising
> Cities are flooding
> Temperatures are soaring
> Forest fires are raging
> Crops are failing
> Common sense is foundering
> Denial is dominating
> The point of no return is looming
> To the few brave enough to face reality
> Welcome to the race against time

The text on all four T-shirts was on the back, with the front having either the website address or an image (created by Ozzy) of the Saltire combined with the book title. The website included an extract from the novel focusing on Mireille, her views on global warming, and the extreme action she takes to resolve it. The website also included photographs of three of the numerous locations in the novel: Rubén looking down onto Princes Street wearing the 'If you're not afraid...' T-shirt, whilst against the backdrop of Salisbury Crags, former European Shokokai Karate Champion Jimmy Pace (who featured in the novel) was wearing the denial T-shirt, and I was wearing the Race Against Time T-shirt at Murder Hill in Gullane Bents.

I decided not to have the first T-shirt on the website as the main focus of it was on the novel.

Things appeared to be moving forwards. Yet I was troubled by the hypocrisy I had encountered from the T-shirt print company. From the onset they'd been pushing organic Stanley Stella Unisex Tees under the premise that they were environmentally friendly when plain cotton Tees were not. The price tag of around £8 more per organic Tee was therefore justified. Except it wasn't. I found out by accident that the price of an organic Stanley Stella T-shirt was only slightly higher (less than £1) than a plain cotton T-shirt of similar quality. When I confronted the print company they had the audacity to use the excuse that they had an agreement with the other T-shirt print companies to sell at this price. The hypocrisy, that they were exploiting global warming to make higher profits, didn't seem to register with them. Perhaps they felt it was okay as all the other T-shirt print companies were doing the same thing. Only when I said I would supply the organic cotton T-shirts myself did they eventually agree to reduce the price. However, when the final bill came in I realized that they had probably increased the print cost to compensate for reducing the cost of the T-shirts, thereby maintaining their outrageously high profit margin.

I was sickened by the hypocrisy of it all. Although the website was up and running, I stopped short of pricing or selling the T-shirts. The third T-shirt was the only one that might have sold, anyway. My intention was to sell at cost price, but for me not to lose money on it, the T-shirt would have to sell at a relatively high price.

The final straw came before a planned demonstration in Glasgow during COP26. Coaches had been arranged to travel through from Edinburgh the morning of the demonstration. I searched online and found the council offices who were organising the coaches, emailed them to tell them I would be visiting early afternoon to drop off approximately 80 T-shirts to hand out to anyone who wanted one. I got a taxi to their offices to find the lights blazing away but the doors locked. I phoned the number of the office I'd emailed earlier, to be told that they'd read my email, but I was at the wrong address as they'd closed that office over two months ago. I struggled to understand how they were clearly supporting the demonstration to save the environment yet had moved location, not bothered to change their address online ... and had left all the lights on. More hypocrisy! The woman I spoke to on the phone took my number and said that someone from the council would be in touch that evening and arrange to pick up the T-shirts from me.

    I didn't believe that would happen, and I was right; no one contacted me that evening.

    Ian took some of the T-shirts, Reno took a batch of them for the guys at his gym, and Rubén took some more for his work colleagues, leaving me with about 100 T-shirts in boxes, where scores of them still remain. After a couple of years, I closed the website. No one had shown an interest in it, and I had lost interest in promoting it. The entire venture had been badly planned on my part. I was now beginning to have doubts that the best way to get a message across was through a fictional story, especially one set several years ago. Yet, despite that, I started work on one additional scene and yet another rewrite of the novel.

    I was clutching at straws now, holding on to the hope that I would find a way to break the Catch through the novel. Deep down I was disillusioned with the hypocrisy I'd encountered. The news seemed to be filled with reports of wildfires in certain countries, and storms and flooding in others, yet all people seemed to care about was how to make more money.

Although I felt the fourth T-shirt was weak the final lines now held an added poignancy for me. *For those brave enough to face reality, welcome to the race against time.* As each day passed I struggled to face the reality of the situation without feeling acutely depressed. Denial now held the promise of sanctuary, an escape from the anguish and fear of the terrible future that awaited the children.

IX

I had a meeting planned with Rubén for some time. Topic of conversation – trees and mass planting thereof. Perhaps it was a forlorn hope, but I felt that if everyone worked together we could plant enough trees to reverse global warming. Rubén dropped by on a Saturday morning. He kicked off the conversation and within the first minute I realized I had been naive again. The longer he spoke, the more complex the topic became. He knew this field inside out, and as soon as a possible question sparked in my conscious mind, he was already halfway through answering it. I had a list of questions I'd prepared, but one by one they became redundant.

To save myself from looking stupid I kept my mouth shut. As much as I didn't want to accept it, the situation looked bleak; no matter how many trees we successfully planted the benefit to the environment would be so slight we would barely notice it. It seemed we were on a path to destruction, and a few million trees wasn't going to divert us from that path.

The entire meeting was depressing. I stayed seated for a long time after Rubén had gone and stared into space. I was numb. The hypocrisy I'd encountered with the company that had done nothing to promote the first video, and the T shirt company's greed and absurdly inflated pricing of organic cotton clothing, coupled with the information provided by Rubén, left me on the brink of despair. I needed to break this train of thought – I grabbed a bottle of water and headed to the gym.

Whilst there I managed to shut out the morning's conversation and work methodically through my short session.

Halfway back I stopped to take a breather. I watched as one of the residents of the flat opposite threw several big carboard boxes into the black bin for general waste. It was pathetic, especially as the paper and carboard recycling bin was right next to it. I wanted to shout at him, ask him why he was being so incredibly stupid. But what was the point - he wasn't going to change. And if he wasn't going to change, why would anything I said, or the team said, make a difference.

I still didn't understand why people behaved in this way. It simply didn't make sense. I had hoped that in writing the novel Mireille might find answers as to why there was such a disparity. Why was a small section of society deeply concerned about the future of the planet, when a far bigger section didn't seem to care? If I could answer that question, then I'd be part of the way to answering the main question - how do we break the Catch? How do we convince everyone that we need extreme measures now and not some time in the future. How do we make the public accept the fact that we're rapidly approaching the point of no return? How do we make people care? If we accomplish that, then in theory at least, we have everyone working together to resolve this crisis.

We needed everyone. We needed unity and we needed it quickly.

So what was it? How could there be such a vast divide between being afraid of the future - a future of failing crops and the food shortage that goes with it - and those who look forward with optimism to a future of prosperity and opportunity.

Were there different types of personality, different attitudes at play here? Did the optimistic automatically dismiss the warnings, or were they just too busy to pay attention. Or did they just think, 'well, what I do will have little or no impact, and it seems no one else is worried, so why should I be?'

And why was I so concerned? I didn't have children. Perhaps it was something to do with the fact that I'd spent the majority of my adult life in law enforcement. Keeping the peace, trying to help those I came into contact with, ensuring they were safe from harm. I also now had more time to think, to ponder, to fret. If I had a busy routine to adhere to, would I be this worried?

I was looking for straightforward answers but there were none. All these questions; I was sick of asking them every day, going over the same ground again and again when there was no black-and-white answer – there were too many variables involved. Too many countries, too many cultures, too many belief systems. Altogether far too much grey.

You can't make people suddenly change the way they think. Change is slow. And in respect of global warming it was going to be too slow. There seemed no way forward. The whole thing was depressing.

The entire mission was pointless.

By now I was outside the entrance to the block of flats. Two steps up to the main door, first step was steep, and I had to focus on it, supporting myself against the wall. Almost there, one more step and in, along the corridor, my hands now shaking as I fumbled with the keys to the flat. Inside, I stood facing the hall mirror.

But I didn't see Ethan Hunt as I usually did. I had been pretending the whole time. Pretending I was someone far younger and in great shape physically when, in fact, I was a wreck. Two organs removed, kidney disease, almost constant gout due to poor kidney function. Then there was the heart failure, doubtless due to the vast amount of water I had to drink every 24 hours. I was a 67-year-old disabled man who wore a splint on a paralysed left foot and needed a walking stick as a result of nerve damage to the left leg. The gym had been a lie; my sessions lasted less than ten minutes. I was working with children's weights, weights that were so light I wouldn't even have warmed up with them when I was younger. With the depression that had set in during the walk back, the false reality that said I could do something to stop global warming had disintegrated. And the other false reality – that I was going to recover my strength, mobility and youth – had fallen away with it.

I lumbered through to the bedroom, sat down on the bed, then collapsed over on to my side. The room began spinning as it usually did after I'd been to the gym. I closed my eyes to free my mind. I had to try to find another path that would take me out of this deep pit of depression that I now found myself at the bottom of.

I needed to end this project before it destroyed what little remained of my will to live. There was no purpose to what I was doing. The novel, yet again, needed more work. In the past I had my two best friends: Robert, and my dear Aunt Georgina, who encouraged me to write. Even if no one else read my books, as long as they did, then nothing else mattered. Five years had passed since they died within two months of each other, and every day I felt that loss.

I could feel myself falling, sinking closer and closer to despair.

All I could hope for now was to find a way out: a way to let go of the truth of what was going to happen to the children and join the billions who were in denial, and enjoy life.

I drifted off to sleep.

Two hours later I opened my eyes and sat upright.

But I was alive!

When I read Banco, the sequel to Papillon, it was the line that struck me more than any other line in any other book. Henri Charrière went through so many adventures after his escape from Devil's Island. He worked hard, and made great sums of cash, only to lose everything again and again through terrible misfortune. But despite all these losses and the knowledge that he had to start all over again, he kept reminding himself that he was alive - each time feeling the euphoria that went with knowing how lucky he was to have survived and escaped from the horrors of Devil's Island.

I knew I was just as lucky to be alive. So it was easy to repeat his words. But not just to repeat them, to mean them as well, to feel how privileged I was to be alive and be able to go to the gym, when so many others couldn't. The gym may have been a lie, in pretending I was young, but it was a good lie. It was a positive lie. I was also privileged to have a team that supported me in taking on the challenge of global warming, even though I now had to accept that we were going to fail.

Fail yes, but not give up. I would be betraying the team if I gave up now, when they were ready to continue. Rubén may see the hopelessness of the situation, but he was still willing to give his time to help. There had to be a small part of him that still believed I might come up with something. Whereas Ian, Ozzy, and Seán were still optimistic. Reno was different. He still had little interest in global warming, but he was there ready to help, no doubt because he felt there may be violence, and if there was, he wasn't going to miss out on it.

The path ahead was clear; I was not going to find a way to break the Catch, but I would finish the novel. I would have liked to have made another video, but one step at a time.

X

I republished the novel in 2022. The plan was to then create a new video to promote it. But my attempt at denying the implications of global warming had been, in great part, successful. I now saw little point in pursuing a second video. That is, until three years later when I watched the documentary *The Year Earth Changed*. The film showed the planet start to heal itself at a surprisingly fast rate during lockdown.

This confirmed my thoughts that we needed extreme measures to prevent the major crops from failing. If the public accepted extreme measures during the covid pandemic, then if the right button was pushed, they would accept similar measures again – and the Catch would be broken.

The documentary had stripped away my hastily built wall of denial. I was now back in touch with reality, which meant I was compelled to write the text for the second video. I decided that this video would apply to the entire planet, and, most of all, it would be greater (in length) and bolder. My goal was to make it clear from the onset that this was about the children, and that the adult generation was, despite the warnings about global warming, failing the children. I wanted to examine the possible reasons for this failure, along with the terrible implications for the future. I felt I needed to touch on the one thing that could prevent that future from unfolding, and the Catch that prevents it from taking place. The purpose of the video was to highlight that we must somehow find a way of breaking the Catch, and in so doing, protect the children.

The one thing that did take time was not the initial writing of the text, but the rewrites. Eventually I managed to pull together a version I was happy with:

When did it happen? What was the precise point? The point where we lost our way? Our purpose, our job, was to survive, and to do that we had to watch over our children. This was, after all, a primordial instinct – to reproduce and protect our offspring. We could fail at everything else, but that took precedence. So how could we let this happen?

It's not as if we weren't warned. The scientists, the experts repeatedly voiced their concerns. They made predictions, and we watched their predictions come true. We saw the changes: the melted icebergs, the dry and cracked land where lakes used to be.

We read about the deaths: from rain so fierce it drowned those it trapped in their basements to fires that raced with the wind whilst overtaking cars and melting tyres, consuming terrified drivers and passengers in the process. Deaths from cyclones, hurricanes, tornadoes ... storms so fierce they ripped buildings from their foundations casting them, and their occupants, into oblivion.

Yet still we continue to contribute to our demise.

Insanely, we stand every summer at the Salt Lakes and pose, smiling for photos alongside the giant temperature gauge showing record highs.

Maybe we feel we can smile. Surely, we are doing enough to resolve the problem. We are recycling, changing our petrol and diesel cars to greener vehicles, spending a minute less in the shower, remembering to switch our lights off...

So we don't feel guilty about going for a hybrid vehicle instead of pure electric, when we should be using public transport. We don't feel guilty when we flock to the airports to fly abroad on our holidays, when we should be spending those holidays at home, volunteering to fight forest fires, plant trees, or build flood walls.

When we should be cutting back on our carbon footprint, we are adding new baby footprints. Why don't we feel guilty about starting a family, or adding to our existing one? Don't we realize the future those children are going to have? A future where, with the seasons so out of synch, the major crops fail. With not enough food to go around, police officers and soldiers are going to break ranks to look after their families ... and that is the point when law and order will start to break down. Smaller tight knit communities may hold out longer but in time they too will fall as a wave of violence sweeps across the globe, unchecked, uncontrollable ... unforgiving.

We should be thinking of what will happen to our children then, before it becomes too late to do something about it.

But we're not thinking about it.

Is it, as Al Gore warned us two decades ago, too inconvenient a truth to handle?

Is it something as simple as denial?

Is it too big, too horrific, for our minds, our sanity, to encompass?

Or are we simply too busy to pay attention? Are we so caught up in the present to consider the future? Are we too busy making sure our offspring have the things we didn't have? Ensuring they are guided down the right path, go to the right schools, make the grades to go to university, aren't led astray, don't become alcoholics or heaven forbid, start taking drugs.

Are we so caught up in these aspects of their upbringing that we miss the most important one?

Or is it because our adult generation hasn't experienced the horrors of a world war to the point where the fight to survive of our ancestors has now become a fight to acquire more: a larger home, bigger car, smarter tv's, better holidays abroad where survival is no longer about life and death but a buzzword used in game shows, and reality television.

Have we, living in a privileged world, become blinkered to real threat, unable to distinguish between false news and hard truths. The line between real life, films and YouTube videos blurred. And in our inability to focus we are going to miss the one key factor, the one thing we should always be watching for on the horizon, and we are going to race past it without noticing ... the point of no return. The point where the momentum of our oceans and planet getting warmer has gathered such force it will continue to accelerate no matter what we do. How can we miss seeing that, the most brutal and cruellest of all truths? We won't see the biggest forest, even when its billion trees are ablaze, burning with such ferocity that the smoke cloud from it blocks out the sun and moon ... blinding an entire planet.

Will the reasons for our failure to protect our children matter then? There will be no history books, there will be no future generations to learn from our mistakes. Perhaps in distant galaxies the young might sit exams on the demise of a species on a planet five billion light years away. They would learn that our leaders met once every 365 days to try to resolve the issue, but those leaders had their own agendas, each with a higher priority than survival. The young students would learn about the lack of unity of the human race – tribes speaking different languages that had continued, for thousands of years, to fight each other over religion or lines on a map.

They would learn about the naivety of that race, who were so caught up in the good life that they forgot to do what was necessary to survive. Concerned with the pursuit of happiness for themselves and their children, they never saw the nightmare future those children would face: soaked, hungry, shivering, taking shelter in caves from the storms and the constant rain, hiding from the gangs that had lost all ties to humanity and were consumed with desire and hunger, looking for their next meal, their next victim. Gangs of murderers and rapists, that searched for our children, unless our children were now in those gangs. And how long would those gangs survive before they consumed each other?

The reasons won't matter then, the could'ves and should'ves. We can't even blame the politicians, for they are us, parents who love their children, and they are in a position to do something, but they don't, we don't, and the few of us that have really tried to, have been ostracised.

The one thing that could save us - getting the leaders of every country in the world to unite. And in their unity introduce the same extreme measures, and back those measures up with the police and the military. The problem being there is a Catch: before leaders will consider taking extreme measures, they have to believe the public will accept those measures. And for that to happen the situation has to worsen considerably. The Catch being that by then we'll have passed the point of no return, and it won't matter how extreme the measures are. Nothing will matter then.

We can call it denial, naivety, stupidity, ignorance, insanity; whatever name we give it the result will be the same. There is no escaping the terrible, unforgiveable, shame of it: we continue to sow the seed, leaving those we love most, our beautiful children, to reap the whirlwind.

Rubén edited it, Ian proofed it, then Seán recorded my rendition of it. I had concerns that I should have asked Ozzy to do the voiceover, but Seán and Ozzy liked my version so things moved ahead despite it being four times longer than the previous video. I had also removed the optimistic ending.

So, the video ended on a pessimistic note, as it should. It had to be honest. At 11 minutes it was far too long, but if viewed from the perspective of the threat to millions of children then 11 minutes was criminally short. But would 11 minutes help bring down the wall of denial? Would it make a difference? Assuming of course that this time it was promoted correctly and reached the target audience.

No. It would probably make no difference whatsoever. To do so the Catch had to be broken. However, I had tried, the team had tried, and we'd failed. It was up to others to break it now.

XI

When I left hospital in 2004 I moved into a rented flat, and I'd lived there for the last 21 years. However, the owner was now selling and I was moving to Falkirk. Both Rubén and Reno had volunteered to help with the move, so the move itself was only going to cost the van rental. The drawback: we would have to wait until after 9 am to pick up the van from the rental garage. On Saturday, the day before the move, on a whim, I phoned big Sean, who owned Bee Smart Removals. I knew Sean from the gym I went to when my training sessions lasted nearly two hours. His company had bent over backwards to help with my temporary move when the flat was undergoing redecoration. Yet again he helped me out, kindly offering to lend me one of his vans. This seemingly minor change, made a significant difference time-wise. The early start allowed me to take both Reno and Rubén for lunch after we'd completed the move and returned to Edinburgh. Reno dropped the van off while Rubén and I headed for the bar restaurant around the corner from my rented flat. I ordered for Reno, who arrived in time for the meal being served.

    Before today Reno and Rubén had met briefly on two occasions. Yet working together on the move, they had bonded, and the conversation flowed between them as if they were old friends. I had been running on adrenaline and there was one thing on my mind; I had to confirm a few things that would allow me to start afresh with my new life in Falkirk. With my handing the keys over and leaving Edinburgh the following morning, I didn't know when I would get another chance to speak to Rubén face-to-face. I waited patiently for the right moment. It came when the three of us had finished eating ...

Looking directly at Rubén, I said, 'Do you remember the conversation we had about the trees?'

Reno looked puzzled. 'Vaguely.'

'I've been thinking over what you said then. Even if we exploited the land on every uninhabited island and sectioned off huge chunks of land in remote areas of Greenland, Russia, Australia, and every country that has vacant land, and cleared those areas completely of people – making them no-go zones and created around them giant firebreaks. I mean sparks and burning timber can only fly with the wind for so long before they burn out ... and if we plant the best trees for that particular region, with the correct spacing between the trees, effectively exploiting every acre of land on the planet ... and if we start making more wooden furniture and stop burning wood and other forms of fuel that store carbon – thereby minimizing the planet's carbon footprint whilst maximizing the growth of, and demand for, carbon storing trees. Even if everyone in every country in the world did this, it wouldn't be enough, would it?'

The scientist sat in silence for a few seconds before answering. 'It would help, in the same way that taking every petrol and diesel car off the road and substituting them with electric cars would. But even that, and everything you've spoken about plus everyone in the world recycling as much as possible, wouldn't be enough.' He paused then added, 'We need extreme measures. Measures the public will resist.'

'So, we're back to the Catch then?' I said.

'What catch?' asked Reno.

I turned to look at my friend of 30 years. Up until now he'd never shown an interest in global warming. I said quietly: 'The solution to the climate crisis is in finding a way to break the Catch – we need extreme measures, but governments and world leaders will never agree to introducing them until the situation becomes so bad that they feel the public will accept them. The Catch being that by then we will have passed the point of no return and it won't matter how extreme the measures are. Nothing will make a difference then.'

'What extreme measures are you talking about?'

'No unnecessary flights, no unnecessary transportation of goods or people between countries. Imagine covid lockdown but instead of a couple of months, we'd be talking years. Possibly a decade. It wouldn't happen overnight; it would be phased in over time, maybe a year. But by the end of that year the airline industry would probably cease to exist.

'Tourism would end; the entire planet would have to readjust to working in a new direction. But it's not even worth thinking about as the measures aren't going to be introduced in time. Only when the major crops fail will things start to happen, but by then it will be too late, and it won't take long before the public realize it, and law-and-order won't just break down – it will end.'

Rubén took his leave at this point. He had friends to meet, and besides he had heard me go on about passing the point of no return too often.

I walked Reno over to his car.

'So when do you see law and order ending?' He asked.

'Depends on when the crops fail. Rubén's parents had a large garden full of almond and olive trees, and every year they yielded enough of both for their entire family and their friends. For generations that took place, but a few years ago, what with the heat and lack of rain, the olive trees were decimated, only a handful surviving, whilst the almond trees were wiped out completely. Crops are failing already, but on such a small scale that it's not news – so it goes unnoticed.'

'So how many years then: ten, twenty?'

'I'd be guessing, but the changes which have been quite slow up to now, will worsen at an increasingly faster pace. More floods, more forest fires, more icebergs disappearing.'

'So the planet's fucked then?'

I nodded.

Reno shook his head. 'When the public realize that, they'll tear world leaders and government officials limb from limb.'

'But the public wouldn't go along with extreme measures anyway.'

'I don't doubt it,' said Reno, 'but they'll still tear them apart.'

He opened the driver's door. I held my arms out at my side, expressing my incredulity. 'But why would they?'

My friend gave me the look that signified I was being naïve again, 'Because when things start falling apart all the public will see is that leaders and governments made an assumption, when they should have given the public the choice.'

I said, 'Well, they'll have trouble finding government officials and world leaders as they will be hiding deep underground in their carefully built, and supplied, secret bunkers.'

Reno laughed out loud, attracting the attention from several people on the street.

'Are you mad!' He said jokingly. He then stopped smiling, 'When the public realizes that the point of no return has been passed, they will see things from a completely different viewpoint. Then, as far as they're concerned governments and world leaders have betrayed them ... and have sentenced their children to a cruel and terrible death ... and it won't matter how well hidden, or how far underground the world leaders' bunkers are, the public, fueled by rage, will find them ... and they will breach those bunkers. God help the politicians and world leaders then.'

I don't remember watching Reno drive off, or how I got back to the flat. I was supposed to be clearing the flat of the rest of my belongings then giving it a good clean prior to handing the keys over. But instead of working I spent the rest of the day and the night lying on the bare mattress looking at the ceiling and thinking about what Reno had said and how it led to the Catch being shattered. I would hand the keys over the next day, unaware that two weeks later I would be sent a bill by the letting agent for close to a grand for leaving stuff in the flat and not having given it a deep clean. Had I known, though, I don't think I would have been able to do anything differently. My mind was totally caught up with the implications of Reno's words, that I couldn't think of anything else. The first thing I needed to understand was how it was that five of us had failed to find an answer after so many years of thinking about it, when Reno had come up with the answer within a few seconds. How was that possible?

When Co-vid hit Scotland, Tesco and other supermarkets would no longer deliver large numbers of six packs of water. This policy did not end when vaccines became available. I needed to drink around eight litres of alkaline sparkling water every 24 hours to fend off urinary tract infections. As soon as he became aware, Reno would pick me up on Sundays and drive me to Tesco stores whereupon we would empty the shelves of sparkling water.

I told an old friend, Lesley, about this. She also knew of Reno's frequent visits to see friends in hospital. She commented about his compassion. She knew him, but clearly she'd never seen him in a fight. Reno wasn't compassionate. He was pragmatic. He came close to fitting the personality of the main characters in two films, *Green Book* and *The*

Intouchables, though he was probably closest to the character of Bud White in *L.A. Confidential*. The main difference being that the violence displayed in these films was tame in comparison to the violence Reno had meted out in his life.

To say he was streetwise would be a massive understatement. When I thought about it, it was obvious he would see through all the grey and come up with an answer. Would it work in practice though?

A referendum for the population of the entire planet would be a logistical nightmare. Even if it was done at the right time, and it was made clear that it would be only done once, there would be no second referendums ... and it would be one count for the entire planet, so that countries couldn't say 'well, we didn't vote for extreme measures so we're not going to implement them.'

Then there was the issue of countries that were predisposed to voting in a certain way: Third World and developing countries, and countries that had minimum carbon emissions yet were suffering the most from global warming.

Not forgetting the millions of workers in the industries that would be hit the most by extreme measures. It would be predictable that those employed in the airline industry and tourism would simply vote against extreme measures. But would they? Would parents who enjoy the good life? Flying their family abroad on holiday twice a year is one thing particularly as so many are doing it. Yet casting a vote, having to make a yes or no decision would be a different matter. In a voting booth you're on your own, and you're put on the spot ... denial is swept away when you have to not just make a decision but take on the present and future responsibility of that decision: are you going to vote against the future of your children, knowing the future those children are likely to face?

When a man or a woman is individually, rather than collectively, put in front of their responsibilities, they are far more likely to act responsibly.

The decision as to whether extreme measures should be introduced should, as it would impact everyone on the planet, be forced upon everyone, and not be taken by a few men and women.

Covid 19, as terrible as it was, had shown us the way forward – that similar measures to those introduced in lockdown would allow the planet to heal itself. But it was a path that meant enduring years of discipline, deprivation, hardship and sacrifice.

With a population in denial, taking that path would not be considered an option.

A referendum for the planet would almost certainly be dismissed out of hand by those in power. They live the good life, and they wouldn't want to give that up. They would be terrified of a referendum knowing it would eliminate the denial that the vast majority of the public appeared to be immersed in.

That referendum was one way to eliminate the denial; the major crops failing, and the horror show that would follow, was the other.

Copyright Acknowledgements

Heller, J. (1961) *Catch-22*. Simon & Schuster

An Inconvenient Truth (2006) [Film] D Guggenheim (Director) USA: Paramount Classics

Stand Up Guys (2012) [Film] F Stevens (Director) USA: Lionsgate

Mission Impossible (1996) [Film] B De Palma (Director) USA: Paramount Pictures

The Avengers (1961-1969) [TV Series] S Newman (Creator) ABC Television

Rich Man, Poor Man (1976) [TV Miniseries] D Greene, B Sagal (Directors) USA: Universal Television

Peter Gunn Theme (1959) [Song] H Mancini

The Blues Brothers (1980) [Film] J Landis (Director) USA: Universal Pictures

Elvira Madigan (1967) [Film] B Widerberg (Director) Sweden: Europa Film

Terror in a Texas Town (1958) [Film] J Lewis (Director) USA United Artists

Irreversible (2002) [Film]. G Noé (Director) France: Mars Distribution

Charrière, H. (1972) *Banco*. France: Robert Laffont

Charrière, H. (1969) *Papillon*. France: Robert Laffont

The Year the Earth Changed (2021) [Film] T Beard (Director) USA: Apple TV+

Green Book (2118) [Film] P Farrelly (Director) USA: Universal Pictures

The Intouchables (2111) [Film] E Toledano, O Nakache (Directors) France: Gaumont

L.A. Confidential (1997) [Film] C Hanson (Director) USA: Warner Bros.

About the Author

A H FitzSimons was born in Johnstone, near Glasgow, in 1958. He served in the British Army in the 1970s and 1980s before joining Lothian and Borders Police. He studied in Edinburgh, winning two Scottish Business Education Council Awards, including the C.A. Oakley gold medal. He began writing in 2005 following a sixteen-month stay in hospital.

Also by A H FitzSimons

Non-fiction
The Fight 2024
The Time Machine

Fiction
The Game
Break Lima
HK9
Not Proven: Fair Game

The video quoted verbatim in Chapter 10 with the title 'What will we tell our children?' should be available on YouTube sometime in January 2026.

www.ingramcontent.com/pod-product-compliance
Lightning Source LLC
Chambersburg PA
CBHW061234070526
44584CB00030B/4113